INVESTMENTS

THE EASY GUIDE TO BUILDING WEALTH WITH AGRICULTURAL BUSINESS FOR BEGINNERS

ALEX NKENCHOR UWAJEH

Legal Disclaimers

All contents copyright © 2016 by **Alex Nkenchor Uwajeh**. All rights reserved. No part of this document or accompanying files may be reproduced or transmitted in any form, electronic or otherwise, by any means without the prior written permission of the publisher.

This book is presented to you for informational purposes only and is not a substitution for any professional advice. The contents herein are based on the views and opinions of the author and all associated contributors.

While every effort has been made by the author and all associated contributors to present accurate and up to date information within this document, it is apparent technologies rapidly change. Therefore, the author and all associated contributors reserve the right to update the contents and information provided herein as these changes progress. The author and/or all associated contributors take no responsibility for any errors or omissions if such discrepancies exist within this document.

The author and all other contributors accept no responsibility for any consequential actions taken, whether monetary, legal, or otherwise, by any and all readers of the materials provided. It is the reader's sole responsibility to seek professional advice before taking any action on their part.

Reader's results will vary based on their skill level and individual perception of the contents herein, and thus no guarantees, monetarily or otherwise, can be made accurately, therefore, no guarantees are made.

CONTENTS

Introduction 5

Building a Profitable Agricultural Business .. 8
   Breaking It Down 9
   Create a Business Plan 10
   Understand Your Market 12
   Seek Business Advice 13
   Choose the Right Legal Structure 13
   Register Your Business 14
   Business Financing 15
   Launch Your Business 17

Making Money in Agribusiness 18
   Broaden Your Debtor Base 19
   Diversifying Income Streams 21

Investing in Agribusiness 25
   Farmland REITs 26
   Agricultural Stocks/Equities 28
   Mutual Funds 29
   Commodity Futures 30
   Exchange Traded Funds 32

The Risks and Rewards 34

Case Study: Making $1 million in
Agribusiness 37
  Case Study 1: Organic Strawberry Farming 37
  Case Study 2: Quail Farming 39
  Case Study 3: Herb Farming 40
  Case Study 4: Aquaculture 41

Conclusion 43

INTRODUCTION

There has been a noticeable increase in the number of farmer's markets right across the country in recent years, which highlights the increased demand for fresh, locally sourced produce.

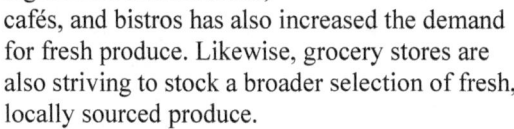

The same rise in popularity for 'farm to table' ingredients in restaurants, cafés, and bistros has also increased the demand for fresh produce. Likewise, grocery stores are also striving to stock a broader selection of fresh, locally sourced produce.

When most people think about where the produce comes from at farmer's markets or large supermarket chains, they usually picture massive, large-scale farms spanning hundreds of acres growing, raising or procuring their agri-products in bulk quantities to keep up with demand.

What you may not realize is that smaller farms and produce suppliers also have a magnificent opportunity to generate huge profits in the agribusiness industry that larger farms just can't compete with.

Large-scale farmers have the benefit of selling mass quantities of their products to large supermarket chains across the country, and even exporting some products overseas to other countries. Unfortunately, large farmers are also at the mercy of weather conditions and changing markets at the same time as trying to maintain control of large operating costs.

By comparison, smaller agribusiness owners have the advantage of being able to supply local demand with a far greater selection of products. There are restaurants, cafés and local grocery stores that would much prefer to source their produce from a local supplier than have it shipped in from another supplier in another state.

Investments: Introduction

No matter what your plan is for your agribusiness, the key to your success lies in creating a strong foundation that will support your business into the future.

Are you ready to get started? Let's go…

BUILDING A PROFITABLE AGRICULTURAL BUSINESS

When most people picture farming, they visualize huge acreage and a long-suffering farmer toiling away with back-breaking labor on the land from dawn until dusk.

What you may not realize is that technology and modern farming equipment has helped to increase the efficiency of many farming practices.

Depending on the type of agricultural business you want to start, you may also be able to start small and scale your business up as profits increase.

In order to build a profitable agricultural business, there are some basic steps you'll need to work through before you begin.

Breaking It Down

The key to building a profitable agricultural business is to break the business structure down into separate components:

- Production of the final product
- Equipment, technology and machinery
- Packaging
- Finance and book-keeping
- Sales and marketing
- Real estate
- Transport and logistics

In order to produce your agricultural product, you will need somewhere to grow, cultivate, or raise it. You'll also need the appropriate equipment and machinery to operate your business effectively.

Once the product is created, it needs to be packaged, prepared, or arranged to be shipped out to its final destination, which also means determining the type of transport, shipping, or other logistics needed to get your final product to your customers.

You also need customers willing to purchase it at a profitable price from you. You'll also need to keep track of your business income and expenditure to ensure it's running profitably.

Each separate component of your agricultural business has an important role to play in your overall success.

Create a Business Plan

Always remember that farming is a business. In order to ensure you have all the elements of your business structured correctly, it's a good idea to sit down and write out a comprehensive business plan.

At the beginning of your business plan, really think about how you would describe the vision you have in your head for your farm. Then write your vision down on paper so others are able to more easily able to understand what you're aiming to achieve.

Your business plan should list all the things you need to buy in order to start your business. When you've created your list, you'll have a much clearer idea of your approximate start-up costs.

From there, you'll also need to understand your estimated ongoing costs. Take the time to explore your options for sourcing any supplies or materials you need to operate your agricultural business and know your costs.

If your products need to be packaged or shipped to your customers, you'll also need to factor in any packaging and shipping costs.

Spend some time seeking out potential customers for your business. In order to generate profits, you'll need paying customers to buy your products.

Research your market carefully and understand what opportunities exist for future expansion as your business grows. Your business plan should also itemize any marketing tactics you intend to use to attract new customers to your products.

Finally, create a plan for how you'll grow your business over time. Your business plan should include some provision for business expansion as profits start to roll in.

Understand Your Market

If you hope to succeed in your agribusiness venture, it's vital that you understand your market.

You'll need to know how customers expect your products to be packaged and presented. You'll also need to research pricing and determine whether the prices you intend to charge will cover your costs.

You need to take the time to reach out to potential customers or distributors. After all, there's no point spending your time or money creating your final product if you don't have customers willing to buy it when it's done.

Even if they do want to buy it, you still need to get your product from your business premises out to your customers.

Take the time to connect with prospective new customers or distributors. Create a plan for how you intend to market, promote, and sell your finished products.

If you're producing crops or livestock, you may qualify for assistance from the Farmer's Market Promotion Program. The program is designed to help develop new market opportunities for farm and ranch operations.

Seek Business Advice

Most states across America have departments willing to offer advice and support to new business owners. For example, you might approach the Farm Service Agency (FSA) and check what services might be available to you.

FSA service centers may offer varying levels of advice and information, but you should have access to information relating to business grants, technical assistance, financial advice, helpful resources, and plenty of other business advice designed to set you on the right path to success.

Choose the Right Legal Structure

Once you've discussed your options and your vision with a business advisory center, it's time to determine the right legal structure of your business.

You should know which form of business ownership will be right for your future business plans, whether you operate under a sole proprietorship or a partnership, or whether you form a Limited Liability Company (LLC), a corporation, an S corporation, or a cooperative.

It's important to remember that the ownership structure you choose could affect your eligibility for some types of programs or grants, so always check with your state's Farm Service Agency before you make your decision.

Register Your Business

When you've sorted through the legal side of things, it's time to register your business. You'll need to register your business name with your own state government and apply for a Tax Identification Number.

Don't forget that you may need to apply for various business licenses and permits before you start operating your business. You'll also need to register for State and Local Taxes.

Even if you plan to operate your business on your own to start with, there may come a time when you'll want to hire employees. Before that time comes you'll want to plan for and take some time to understand your responsibilities as an employer. Because it's important that you know the correct steps you need to take when you're hiring staff. Knowing this will save you money, potential hassles and time.

Business Financing

Perhaps the two biggest challenges for any new agribusiness owner is accessing the land and obtaining finance.

The amount of finance you'll need to get started will depend heavily on the type of agribusiness you plan to set up. Your plan might be to start out small with a few basic pieces of equipment on leased land, with plans to add more equipment and machinery as you start making profits. In this case, you might finance your start-up with some of your own savings and perhaps a small loan.

Alternatively, your plan might be to borrow a larger chunk of money to purchase land, stock and equipment to establish a bigger enterprise right away.

No matter what your intention is, you will need to work out your financing options.

It's a good idea to check with the United States Department of Agriculture (USDA) to see if they can assist you with this step. They'll provide you with information and resources that could help you get the financing you need.

In some cases, you may qualify for financing through the Small Business Administration (SBA), or even for land contract guarantees to help you obtain the necessary land to start your enterprise.

You may also be eligible for some financial assistance or even grants. You'll find a list of grants, loans, and other support services here: http://www.usda.gov/wps/portal/usda/knowyourfarmer?navid=kyf-grants

The USDA offers a Direct Farm Ownership Program, as well as a Value Added Producer Grants Program. There are also plenty of other programs available, designed to help farmers and agribusiness owners. For example, the Environmental Quality Incentives Program offers financial and technical assistance that helps farmers address and deliver environmental benefits.

The Rural Energy for America Program (REAP) may also provide grants for agricultural businesses that want to buy or install energy-efficient measures, such as solar panels.

Keep in mind that your fledgling business will also need sufficient operating capital put aside – or spare cash to cover your operating costs until you start making profits.

Launch Your Business

When you've sought advice and created a strong foundation of planning and preparation to work from, it's time to get your fledgling business off the ground.

If you've completed your business plan properly, you'll already have a clear idea of the tasks and chores you must complete each day as part of your duties in operating your business.

Once your business is up and running, you're likely to spend much of your time and focus just maintaining daily operations. That's completely normal.

After all, it takes time and consistency to build a solid business in any industry.

Just keep in the back of your mind that once your business is up and running, there are still plenty of things you can do to expand your business and boost your profits.

MAKING MONEY IN AGRIBUSINESS

One of the biggest challenges facing new farmers, ranchers or agribusiness owners is finding consistent markets to sell their products. If you're new to the agriculture business, it can be even more challenging to try and break into already-established markets.

There's little point in going to all the effort of producing your final product if you don't have customers waiting to purchase it from you when the produce is ready for the market.

A big part of your business plan should have included your marketing plan. You should already know exactly how you'll package, present, and promote your primary product.

Even if you have steady, reliable customers who always buy everything you produce, there are some additional ways you might be able to make your agribusiness more profitable overall.

Broaden Your Debtor Base

Can you imagine if you received a consistent, ongoing contract to supply one large supermarket chain across the country with your product? If you're like the majority of people, you'll be ecstatic by the thought that your agribusiness is guaranteed sales and profits for years to come.

However, if your single large client happens to stop ordering from you without any prior notice, what would happen to your business?

Ideally, it's wise to ensure you don't receive more than 50% of your total business revenue from a single source. The objective is to seek out additional customers and contracts to buy your products in order to make up the other 50% of your total income.

In the event that you lose your largest client, your business doesn't come crashing to a sudden stop, as you still have revenue coming in from your other customers.

In business terms, this is known as 'broadening your debtor base'.

The biggest challenge many business owners face when trying to develop a broader debtor base is that their largest client often increases the amount being ordered.

When this happens, you have two hurdles to overcome. The first is keeping up with production levels to supply the increase in demand. The second is finding ways to increase your other customer orders so that your largest customer doesn't exceed your 50% of total revenue threshold.

The benefit of maintaining a strict policy of not allowing one single client to exceed 50% of your total revenue is that you have the opportunity to keep earning more profits. If your major client increases the number of orders, you will need to continue increasing production to keep up with increased demand, which results in a boost in profits.

At the same time, if your primary client goes over the preferred 50% revenue amount, you'll spend time sourcing new markets and new customers to keep the numbers balanced. The result is more orders, more sales, and more revenue for your business overall.

Diversifying Income Streams

Your agribusiness may take all of your focus and effort to produce your primary product in order to generate profits for your business.

However, if you've chosen a seasonal product, there will be times throughout the year when you aren't selling anything. The result is that you'll have no income coming into the business until harvest time. Seasonal agribusinesses have two choices. They can:

a.) Ensure their pricing and production is high enough to sustain an entire year's operating costs and income from one massive seasonal influx of income, or;

b.) Work on finding additional or supplemental income streams to keep revenue churning all year round.

In order to keep income rolling into your business, it's important to ensure you incorporate ways to generate steady income that keep your profits stable. Essentially, you may want to consider ways to diversify your income streams.

For example, if you grow cucumbers, you might also invest in the machinery and equipment needed to turn your cucumbers into pickles. Your primary profits may be derived from growing and selling fresh cucumbers, but a secondary product can be sold via different distribution channels to retailers to help sustain your business's revenue all year round.

Likewise, if you've chosen to farm quail birds to sell the meat to restaurants or grocery stores, you have the option of selling quail eggs as a secondary form of income into your business.

Only you can determine the right way to diversify your business's income based on the same products you already created. Be creative and consider your options carefully. You just might have more income opportunities in your business than you expect.

If you're unsure how you might expand your market opportunities, contact the USDA and see whether you qualify for the Value-Added Producer Grant program. This program is designed to help farmers develop new product and uncover new marketing opportunities that help to increase agribusiness income overall.

Supplying the Supplier

While agribusiness is almost always associated with the actual farmer producing the final product, there are also highly successful agricultural businesses out there that don't do any farming at all.

Instead, their profits come from supplying the things farmers need in order to continue production effectively.

For example, crop farmers always need a good supply of fertilizer and compost, as well as a reliable source of crop seeds. Farming animals means the farmer always needs a constant supply of food for the number of animals on the farm.

Some crop farmers rely heavily on local apiarists – or bee keepers – to pollinate certain crops for them. Aside from assisting other farmers with pollination needs, bee keepers can also package and sell the honey they produce.

Bookkeepers have the opportunity to manage and control an entire farm's finances through their business, saving the farmer time and ensuring operations continue running smoothly.

Trucking, transport, packing and shipping companies compete for business from farmers, who need to get their produce packaged and distributed to customers quickly and effectively. Farmers also need equipment and machinery to operate their farms, all of which must be sourced from a specialist agribusiness supplier somewhere.

There are a myriad of different suppliers out there generating excellent business profits by supplying the farming industry with the things they need. While they may not be farmers in their own right, they are still a crucial part of the agriculture industry.

INVESTING IN AGRIBUSINESS

So far we've looked at options for starting and operating your own successful and profitable agribusiness. However, there are lots of ways to generate wealth with agribusiness without buying, owning or operating your own farm.

There are plenty of alternatives available to you if you're keen to invest in the farming sector.

Many investors diversify their portfolios with farming or agricultural stocks. It's often seen as a smart strategic move, simply because people will still need to buy food even in difficult economic times.

Historically, agricultural stocks have also achieved steady returns year after year, even in difficult economic times, so they're considered by many to be a safer investment than stocks and other investment types across many other sectors.

Here are some examples of ways you can invest in agriculture and agribusinesses without buying a farm:

Farmland REITs

Investing in farming real estate investment trusts (REITs) allows you to invest in a variety of different farming properties without actually buying the farm.

Essentially, the managers of REITs purchase farmland in a variety of areas and then lease the land back to farmers. The REIT generates income from the rent payments received from various farmers who have leased the land within the trust.

According to data released from the National Agricultural Statistics Service (NASS), U.S. farmland offered investors an average annual return of 12% over the last 20 years. By comparison, the average returns on the S&P 500 for the same period of time were just 9%.

One of the key benefits of investing in Farmland REITs is that your trust might have interests in various farms across several different states. Farmland REITs are also well-known for providing steady returns with low volatility.

REITs give you the flexibility to get your money out again quickly. The market is significantly more liquid than trying to sell a plot of farmland in a regional or rural area.

You have the freedom to choose which particular REIT you invest your money into. As an example, the REIT Gladstone Land (NASADAQ: LAND) began by investing in farmland growing fruit, vegetables, and berries. By comparison, the REIT Farmland Partners (NYSE: FPI) bought land where the farmers were growing wheat, soybeans, corn, and cotton.

When a REIT realizes a profit, it can choose to share the profits with each individual investor in the form of dividends. If you're seeking ways to earn passive income from your investments, it's possible to find REITs paying around 5% dividend yield, which is a far healthier return on your investment than the interest you could earn on the same money in the bank.

Agricultural Stocks / Equities

Instead of investing in real estate trusts, you do have the option of investing in stocks of individual agricultural companies.

For example, you might choose to invest in a company such as Fresh Del Monte Produce Inc. (NYSE: FDP) that focuses on planting, growing and harvesting crops.

Alternatively you might invest in a company such as Adecoagro S.A.(NYSE: AGRO), which derives its profits from planting, growing and harvesting crops, as well as maintaining an involvement in distribution, processing and packaging of agricultural products.

There are also companies that buy and hold farming land. The companies are publicly listed on the stock exchange and the stocks trade very much like any other commodity.

For example, you might buy stocks in the company the Texas Pacific Land Trust (NYSE:

TPL) or Alico (NASDAQ: ALCO) to add to your investment portfolio. The companies invest in buying farmland and you invest in the company by buying stocks.

Alternatively, there are also publicly listed companies that make their profits by serving the agricultural industry, such as Agrium that produces agricultural chemicals, or agricultural equipment manufacturers like Deere, or even large processors like Tyson Foods Inc.

Just as with REITs, many companies will also issue dividends to investors based on the profits generated. The returns can vary greatly, depending on the stocks you choose to invest in, so be sure to do your research and have a specific exit strategy in mind before entering into any investment position.

Mutual Funds

It is possible to invest in mutual funds that focus solely on investing in the agricultural industries. Mutual funds give you a broader exposure to a range of different farming and agricultural businesses.

However, it's a good idea to determine whether the fund you're thinking of getting into is focused on investing in agricultural businesses, farming industry companies, supporting industry companies, or in commodities.

Many mutual funds may spread their exposure to various sectors of the agricultural industry in an effort to mitigate risk and stabilize returns. Others may choose to invest strongly in one particular sector or type of investment over another.

If you're keen to invest purely in the farming industry or a particular aspect of the agricultural industry, spend a bit of time researching your options and returns carefully.

It's also wise to take the time to research each individual mutual fund's fees and past performance before making an investment decision.

Commodity Futures

Investing in futures is not often recommended for the average investor, as they are quite a complex financial derivative product. However, if you already understand how futures contracts are written, you can use your knowledge to invest in the agricultural commodities market.

Put simply, futures contracts are designed to reduce some of the risks farmers and agricultural producers face. After all, the farmer is putting in all the hard work of plowing, sowing, harvesting, sorting, and shipping before they actually receive any profits.

An example of investing in futures is to invest in corn. Imagine if the farmer expected a certain price for each bushel of corn before he began sowing the fields.

However, by the time the corn was harvested the market price for a bushel of corn may have increased dramatically, representing a massive profit for the farmer. Alternatively, the price for a bushel of corn may also have plummeted in the time it took to produce the crop, which could represent a financial loss overall.

The commodity futures market is designed to help alleviate the potential risks farmers face with fluctuating prices.

Your contract is written to represent 5,000 bushels of corn, or 127 metric tons. The actual symbol used on the exchange will change depending on when the futures contract expires.

Exchange Traded Funds

Exchange Traded Funds (ETFs) allow average investors to speculate on the commodities futures market without the same level of risk. ETFs are traded on the stock exchange, just like regular stocks. They're also somewhat like mutual funds, but instead of investing in farming real estate like REITs, ETFs invest in agribusinesses and commodities.

For example, the ETF DB Agriculture Fund (DBA) specializes in investing in futures contracts on things like sugar, coffee, soybeans, cocoa, and corn. There are ETFs out there that invest in exclusive areas of the agricultural industry, such as investing only in grains, or only in live cattle.

Alternatively, there are also ETFs that give investors access to more diversified businesses, such as investing in companies that earn at least half of their revenues from agricultural industries. The remaining portion of their revenues may be derived from various other industries to help stabilize returns and mitigate risks.

As with mutual funds, it's important to carefully consider any fees and check the fund's past performance before making an investment decision.

> Disclosure: I have no position in any of the stocks or REITs mentioned in this book. The companies and trusts mentioned are for illustrative purposes only.

THE RISKS AND REWARDS

Every business enterprise has its share of risks – no matter what industry you're getting into.

Some agribusinesses are susceptible to weather conditions and seasonal factors. Others are prone to legal risks.

Some may be prone to animal illness or disease, while others may have a greater risk of crop or soil problems. There is also the risk of doing business in fluctuating market conditions.

There's also the financial risk to consider. Investing in starting and operating a business can require a significant amount of money to get off the ground.

In order to effectively operate any business, it's important you understand the inherent risks that could affect your enterprise. Know what things could go wrong and find ways to mitigate those risks wherever possible.

In some cases, you might be able to apply for insurance for your property, business, and products. In other cases, it might be wiser to create business operating procedures that help you reduce your risks.

While you might focus carefully on finding solutions to any potential risks that could arise in your business, it's also nice to focus on the rewards too.

Just imagine if your fledgling little agribusiness grew and expanded into a multi-million dollar enterprise. It's certainly possible and there are plenty of awesome success stories out there of regular people doing exactly that by following their passion.

Likewise, every investment also contains an inherent risk. If your goal is to build wealth by investing in agribusiness, it's wise to create a solid investment strategy and don't allow emotions to get in the way of making good investment decisions.

Be sure whether you want to invest for dividend yields to generate passive income, or whether you prefer a speculative style of investing and focus on day trading in agricultural or farming stocks.

Of course, despite the risks of investing in the stock market, or in REITs, ETFs, mutual funds, or the commodity futures market, there is also a great potential to generate wealth from the agribusiness industry without the hard work and back-breaking labor of owning and operating a farm.

There's also the big benefit of being able to get started with your investment goals with a minimal capital outlay. You still have the benefit of investing in the agriculture industry, but you don't have the financial concern of purchasing entire farms, equipment, machinery and stock.

Before you take the leap into either option, it's wise to weigh up all the potential risks and rewards associated with your decision. Look for ways you might potentially reduce or mitigate the risks involved. Then look for ways you might improve your chances of reaping the rewards.

When you're prepared for any eventuality, it becomes easier to weather uncertain economic situations and market volatility without losing your initial investment completely.

CASE STUDY: MAKING $1 MILLION IN AGRIBUSINESS

Before we get into case studies about the reality of making $1 million in agribusiness, it's important to consider the size and scale of the type of business you want to begin.

After all, if you go out and purchase 100,000 acres of grazing land, purchased equipment and machinery, and bought several thousand head of prime cattle, you're likely to have already spent far more than just $1 million getting your agricultural business started.

For the purpose of these case studies, we're going to focus on agricultural businesses that began as small-scale operations and built up to thriving, successful enterprises generating millions of dollars in profits.

Case Study 1: Organic Strawberry Farming

Jim Cochran always loved fresh strawberries, but he found it difficult to find the delicious, fresh, organic fruit he remembered from his grandmother's yard from when he was a child. Instead, grocery stores and supermarkets were selling mass-produced commercially-grown strawberries that lacked the home-grown flavor he missed so much.

He also noticed that many commercial growers use chemical pesticides, fungicides and herbicides to control bugs, disease, and weeds, which posed a potential health problem.

So he decided to become California's first commercial organic strawberry grower. Cochran began his operation by leasing some land near Santa Cruz and experimenting with ways to produce healthy crops without using toxic chemicals.

Today, Cochran's strawberry farm spans 40 acres and produces organic strawberries on a large scale each year. Aside from supplying local restaurants, farmer's markets, and grocery stores, his farm also features a 'pick-your-own' component that encourages people to wander the rows of strawberries at their leisure and pick their own to eat while they're picking or to take home with them.

It's estimated that only 4% of commercial strawberry growers in America are organic, so his business is set to keep growing into the future.

Case Study 2: Quail Farming

Bill Odom originally started raising quail as a hobby to train his hunting dogs. In the off season, he began selling any excess quail to friends and family.

It didn't take him long to realize that there was a blossoming market for quail in restaurants and grocery stores. So he founded Manchester Farms and began raising quail to supply the local market.

Since that time, the farm has grown and expanded in an effort to meet the increasing demand for quail to retailers and chefs right across the country. The farm now produces more than 80,000 quail per week, all of which are grown hormone-free.

Quail farming is a relatively low-cost farming operation to start, and the birds can be farmed for the meat as well as the eggs. They're much smaller birds than chickens, so they don't need as much space to farm effectively.

Quail also have a very short turnaround production time, which allows quail farmers to reap profits in as little as six weeks, as compared to chickens that can take up to six months for a farmer to realize profits.

Quail meat and quail eggs are big in Mexican cuisine. They're also growing in popularity in fine dining restaurants. Quail is also very popular in Vietnamese and Chinese cuisine, so demand for quality quail produce is still increasing right across the country.

Case Study 3: Herb Farming

Roger Sego and his wife Helen started a small organic herb farm near La Center in Washington with the intention of supplying organically grown medicinal herbs to the market.

The farm started out producing organic goldenseal, which is used to treat sore throats, skin problems and digestive problems.

As the popularity of his herbs grew, Sego added more fields to expand their crops, spreading further across their 55 acre farm and added a processing plant on the property to ensure the quality of his products.

He also expanded his product range to include growing organic ginkgo biloba, echinacea angustifolia, and ginseng.

At first he marketed his organically grown herbs to companies who turn it into herbal medicine products to be sold in health food stores. However, as business grew he sought out new markets and added more distribution channels so his products are now shipped all over the world.

The farm also now has its own retail store, further increasing revenue.

Case Study 4: Aquaculture

Andrew Grant had a vision for farming freshwater crayfish – or 'yabbies' as they're called in Australia. His yabby farm concept began with a few small tanks to produce quality yabbies that he was selling to the local restaurant industry across the city of Melbourne.

As profits grew, he expanded the number of tanks he operated and diversified his business to include growing barramundi. It didn't take him long until the farm was producing 50 tons of live fish to the local restaurant industry each year.

Once again, Andrew re-invested his profits into building up his farming capacity and added rock lobster, snow crabs, abalone, and eels to his product lines.

While his aqua-farming business may have started on a small scale, he now operates a highly successful business that generates a seven figure revenue each year. Andrew understood the necessity of diversifying his product lines to keep profits consistent.

CONCLUSION

The agricultural industry could be considered the perfect recession-proof business. Even in difficult economic times, people still need to buy food.

There's also the increase in demand for fresh, locally sourced produce to take into consideration. Consumers want to know that their fruit, vegetables, seafood and meat come from local farms and producers in their immediate vicinity.

Agribusiness is the ideal means of wealth creation, whether through investing or through creating new businesses that generate wealth for the business owners and investors alike.

Creating a profitable agricultural business not only builds wealth for you and your family. It also has the potential to build wealth for the surrounding community.

As your business grows, it provides employment opportunities. Your employees are likely to spend money in your community on food and other local products. Local grocery stores and restaurants thrive and respond to the increase in demand by ordering more from your agribusiness, which results in more growth, more employment opportunities, and more profits for your business overall.

If you're ready to leave the corporate rat-race and take the leap into being your own boss, agribusiness is an excellent option. Of course, there is some work and risk involved in getting started, but the rewards can certainly outweigh the risks.

Take the time to plan your business foundation and structure correctly, diversify your income opportunities, broaden your debtor base, and focus on producing the best possible products you can. The result should be a sustainable business that continues to generate wealth for you and your family for years – or even generations – to come.

Other Available Books:

- In The Pursuit of Wisdom: The Principal Thing

- **Investing in Gold and Silver Bullion - The Ultimate Safe Haven Investments**

- Nigerian Stock Market Investment: 2 Books with Bonus Content

- **The Dividend Millionaire: Investing for Income and Winning in the Stock Market**

- Economic Crisis: Surviving Global Currency Collapse - Safeguard Your Financial Future with Silver and Gold

- **Passionate about Stock Investing: The Quick Guide to Investing in the Stock Market**

Guide to Investing in the Nigerian Stock Market

- **Building Wealth with Dividend Stocks in the Nigerian Stock Market (Dividends - Stocks Secret Weapon)**

- Bitcoin and Digital Currency for Beginners: The Basic Little Guide

- Child Millionaire: Stock Market Investing for Beginners - How to Build Wealth the Smart Way for Your Child

- Christian Living: 2 Books with Bonus Content

- **Beginners Quick Guide to Passive Income: Learn Proven Ways to Earn Extra Income in the Cyber World**

- Taming the Tongue: The Power of Spoken Words

- **The Power of Positive Affirmations: Each Day a New Beginning**

- The Real Estate Millionaire: Beginners Quick Start Guide to Investing In Properties and Learn How to Achieve Financial Freedom

- **Business: How to Quickly Make Real Money - Effective Methods to Make More Money: Easy and Proven Business**

Strategies for Beginners to Earn Even More Money in Your Spare Time

- Money: Think Outside the Cube: 2-Book Money Making Boxed Set Bundle Strategies

- Marketing: The Beginners Guide to Making Money Online with Social Media for Small Businesses

If you would like to share this book with another person, please purchase an additional copy for each recipient. Thank you for your support and thanks for reading this book.

www.ingramcontent.com/pod-product-compliance
Lightning Source LLC
Chambersburg PA
CBHW070413190526
45169CB00003B/1237